STIFLING INNOVATION
DEVELOPING TOMORROW'S LEADERS TODAY

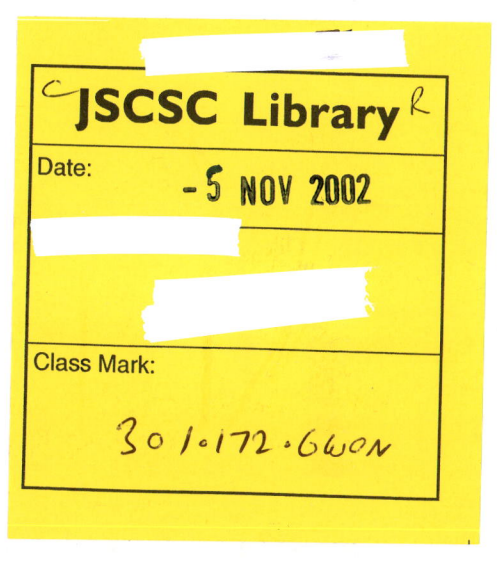

Leonard Wong

Books Express
Specialist suppliers of Military History and Defence Studies
P.O. Box 10, Saffron Walden, Essex, CB11 4EW. U.K.
Tel: 01799 513726. Fax: 01799 513248
info@books-express.co.uk / www.books-express.co.uk

April 2002

*****

The views expressed in this report are those of the author and do not necessarily reflect the official policy or position of the Department of the Army, the Department of Defense, or the U.S. Government. This report is cleared for public release; distribution is unlimited.

*****

Most of this monograph is based on research by the Company Commander's Task Force—group of students from the class of 2001 at the U.S. Army War College. This Task Force, chartered by the Chief of Staff of the Army, was comprised of Colonel Mark Stevens, Infantry; Lieutenant Colonel Don Phillips, Infantry; Lieutenant Colonel Russell Gold, Armor; Lieutenant Colonel Billy West, Field Artillery, Army National Guard; Lieutenant Colonel Timothy Cornett, Aviation; Lieutenant Colonel Patrick Plourd, Aviation; Lieutenant Colonel Richard Kenney, Signal Corps; Lieutenant Colonel Nathaniel Smith, Signal Corps; Lieutenant Colonel Janice Berry, Quartermaster Corps; Lieutenant Colonel Debra Broadwater, Transportation Corps, U.S. Army Reserve; and Leonard Wong, Ph.D., Faculty Advisor.

*****

Comments pertaining to this report are invited and should be forwarded to: Director, Strategic Studies Institute, U.S. Army War College, 122 Forbes Ave., Carlisle, PA 17013-5244. Comments also may be conveyed directly to the author. Dr. Wong can be contacted at (717) 245-3010 or via e-mail at Leonard.Wong@carlisle.army.mil. Copies of this report may be obtained from the Publications Office by calling (717) 245-4133, FAX (717) 245-3820, or via the Internet at Rita.Rummel@carlisle. army.mil

*****

Most 1993, 1994, and all later Strategic Studies Institute (SSI) monographs are available on the SSI Homepage for electronic dissemination. SSI's Homepage address is: http://www.carlisle.army.mil/usassi/welcome.htm

*****

The Strategic Studies Institute publishes a monthly e-mail newsletter to update the national security community on the research of our analysts, recent and forthcoming publications, and upcoming conferences sponsored by the Institute. Each newsletter also provides a strategic commentary by one of our research analysts. If you are interested in receiving this newsletter, please let us know by e-mail at outreach@carlisle.army.mil or by calling (717) 245-3133.

ISBN 1-58487-064-8

FOREWORD

The Army is moving rapidly on the road to transformation. New vehicles are being fielded, doctrine is being written, and alternative force structures are being tested. A key part of this process is the transformation of the human dimension of the Army. The future leaders of the transformed Army will have to be innovative leaders who can operate in a rapidly changing environment in the absence of detailed guidance from higher headquarters.

Dr. Leonard Wong examines how, despite the need to develop and grow innovative leaders for the Army's future forces, the current system in the Army seems to be working against that vision. He argues that current levels of directed training events, dictated training procedures, and disruptions originating from higher echelons are having a detrimental effect on the development of innovation in today's company commanders.

Some may find it difficult to accept the message of this monograph as they point to the many operational and training successes of the Army. However, success today often masks the need for change in preparation for tomorrow's battles. The recommendations made in this monograph need to be carefully considered as the Army continues along the path to transformation. If not, the transformed force may find itself without the innovative soldiers needed to lead it.

DOUGLAS C. LOVELACE, JR.
Director
Strategic Studies Institute

BIOGRAPHICAL SKETCH OF THE AUTHOR

LEONARD WONG joined the Strategic Studies Institute in 2000 after a career in the U.S. Army. His time in the Army included teaching leadership at the U.S. Military Academy, serving as a manpower analyst in the Office of the Chief of Staff of the Army, serving as a strategic planning analyst in the Office of the Deputy Chief of Staff for Personnel, and serving as the Director of the Office of Economic and Manpower Analysis. Dr. Wong has written extensively on Army organizational issues such as downsizing, leadership, and junior officer retention. He is a Professional Engineer and holds a B.S. from the U.S. Military Academy, and an M.S.B.A. and Ph.D. from Texas Tech University.

SUMMARY

This monograph examines the current company commander experience and concludes that the Army values innovation in its rhetoric, but the reality is that junior officers are seldom given opportunities to be innovative in planning training; to make decisions; or to fail, learn, and try again. This controlling, centralized environment results from three main factors. First, higher echelons increasingly are directing training requirements, taking away the discretion of company commanders to plan their own training. Second, higher headquarters increasingly are dictating how training should be conducted, taking away the initiative of company commanders when executing training. Finally, senior commanders increasingly are disrupting training with administrative requirements and taskings, taking away the predictability of company command.

If the transformed Army will require leaders who can operate independently in the absence of close supervision, the current leader development experience of company command will have to change. Consequently, the author asks for senior leaders not to do more, but to do less and thus give subordinates more freedom to innovate.

STIFLING INNOVATION: DEVELOPING TOMORROW'S LEADERS TODAY

> Building a 21st century military will require more than new weapons. It will also require a renewed spirit of innovation in our officer corps. We cannot transform our military using old weapons and old plans. Nor can we do it with an old bureaucratic mindset that frustrates the creativity and entrepreneurship that a 21st century military will need.
>
> President George W. Bush
> U.S. Naval Academy commencement address,
> May 2001

> The development of bold, innovative leaders of character and competence is fundamental to the long-term health of the Army. We must grow leaders, NCOs, officers, and civilians for the future by providing appropriate opportunities for the development of those skilled in the profession of warfighting.
>
> General Eric K. Shinseki
> Intent of the Chief of Staff of the Army,
> June 1999

> **Interviewer**: Do you feel you're being trained to be a creative, innovative and adaptive leader?
>
> **Company Commander:** They're not telling me, "Here, you've got ten crews—train them." They're not allowing me to devise the methods and the ways to get there. They're giving me the egg and telling me how to suck it.

Introduction.

In October of 1999, the Army senior leadership unveiled plans for the Army's transformation. Almost immediately, the vision of deploying a combat-capable brigade anywhere in the world within 96 hours, a full division in 120 hours, and five divisions on the ground within 30 days grabbed the attention of the force and the public. Transformation became the topic *du jour* in the Army. Debate over tracks versus wheels, light versus heavy, and Objective versus Legacy dominated briefings, conferences, e-mails, and letters to the editor. Absent from the debate, but included in at least one slide of almost every transformation briefing, was the call for a transformation in the *human dimension* of the Army. Transforming the human dimension implies that soldiers, and especially leaders, of the future force need to be more flexible, adaptable, creative, and innovative. The leaders in the future brigade combat teams may have to receive the details of their mission en route to their location, put together an ad hoc task force on the fly, or operate for long periods of time in the absence of guidance or supervision.

Future forces will require leaders "to use initiative within intent—leaders who can create cohesive units that thrive in high-tempo, dispersed operations."[1] Future leaders will need to be independent and creative as they craft a plan based on the commander's intent and alter the plan as conditions change. In the absence of detailed guidance from above, these leaders will learn to live with uncertainty, take bold risks, and assume greater responsibility for decisions concerning their unit. In the words of the Army Training and Leader Development Panel, future leaders must demonstrate the competence of *adaptability*.[2] Descriptors of future Army leaders found in transformation brochures commonly include adjectives such as "responsive," "agile," and "versatile." Clearly, the future leader must be someone who can confidently create something new out of nothing based on the needs of the mission.

When references are made to the leaders of the transformed Army, it is common to picture them as mythical "Starship Troopers"—after all, the year 2015 sounds so distant. In reality, those future leaders are today's junior officers—ranging from newly-commissioned lieutenants to today's company commanders. With that realization in mind, how is the Army doing in developing junior officers today to be the innovative leaders of tomorrow? Is anything being done differently to produce a different type of leader?

Unfortunately, behind the seemingly ubiquitous consensus on both the importance of the human dimension in the future and the need for transforming it, a serious disconnect remains between current leader development practices and the type of leaders required by the future force. Put bluntly, the Army is relying on a leader development system that encourages reactive instead of proactive thought, compliance instead of creativity, and adherence instead of audacity. Junior officers, especially company commanders, are seldom given opportunities to be innovative; to make decisions; or to fail, learn, and try again. Senior leaders have, as retired General Wesley Clark suggests, ". . . gone too far in over-planning, over-prescribing, and over-controlling."[3] This move to over-centralized control results in an environment where, as retired General Frederick Kroesen notes, "Initiative is stymied, and decision-making is replaced by waiting to be told."[4] According to Kroesen, "There is no more effective way to destroy the leadership potential of young officers and noncommissioned officers than to deny them opportunities to make decisions appropriate for their assignments."

To use a culinary example, *cooks* are quite adept at carrying out a recipe. While there is a small degree of artistic license that goes into preparing a meal, the recipe drives the action—not the cook. *Chefs*, on the other hand, look at the ingredients available to them and create a meal. The success of the meal comes from the creativity of the chef—not the recipe. In a large hierarchical army, many "cooks" are needed—leaders who can be counted on to follow doctrine competently in their part of the hierarchy. But the environment of the Objective Force calls for "chefs"—leaders capable of operating outside of established doctrine and existing hierarchy.

The Next Generation of Leaders.

The current condition of the leader development process is enough to justify a call to action, but there is another factor that adds to the imperative for change. Newly-commissioned second lieutenants are increasingly coming from the Millennial Generation (sometimes called Generation Y, the Nexter Generation, or the Echo Boomers). Unlike the skeptical, pragmatic, independent Generation X officers moving through the ranks,[5] the next generation is turning out to be radically different. To many, these new Millennial officers represent a breath of fresh air as they differ from their Generation X predecessors in four key areas:[6]

First, compared to Gen X officers, Millennial officers are not pessimists. Unlike the dark, cynical roots of Generation X, this Millennial generation is largely upbeat. Nine in ten describe themselves as "happy," "confident," and "positive." According to a 1997 Gallup survey, 82 percent of Millennial teenagers described their home life as "wonderful" or "good."

Second, Millennials are not self-absorbed. Instead, they are team players. According to a Roper survey conducted in 1998 to rank the major problems facing America today, Millennials aged 12 to 19 named *selfishness* as the top concern more than any other issue. From trends ranging from school uniforms (which were endorsed in the 1996 State of the Union address) to volunteering for community service such as Habitat for Humanity (75 percent of college freshmen reported doing volunteer community work), the Millennials are drawn to group activity. Similarly, the number of applications to the Peace Corps is up 56 percent from last year; applications to AmeriCorps increased nearly 50 percent.[7]

Third, Millennials are not distrustful. In contrast to Generation X, this generation gets along with authority. In a 1997 Gallup survey, 96 percent of teenagers said they got along with their parents. In the 1998 Roper survey of teenagers, in addition to selfishness being the most frequently cited concern facing America, the other top four concerns were: people who don't respect the law and the authorities, wrongdoing by politicians, lack of parental discipline, and courts that care too much about criminals' rights. Contrast this concern for more respect for authority and stricter parental discipline with 1974 poll results showing 40 percent of teenagers reporting that they would be "better off not living with their parents." Millennials are the first generation whose after college plans increasingly include moving back home with Mom and Dad.

Finally, the Millennial generation was not neglected like the last generation. Instead, they are the most watched-over generational cohort in American history. Adults have increasingly been ordering the life of American youth with structure and supervision, ranging from bike helmets to playdates. From 1981 to 1997, the amount of time children aged 3 to 12 spent playing indoors went down by 16 percent. The amount of time watching TV decreased by 23 percent. The amount of time spent studying increased by 20 percent and amount of time in organized sports rose by 27 percent. In 1981, the average 6- to 8-year-old was doing 52 minutes of homework a week. By 1997 it had doubled to more than 2 hours a week. During this timeframe, free or unsupervised time in the typical preteen's day shrank by 37 percent. As a result, this generation is accustomed to being watched, controlled, and scrutinized. To this cohort, the increased security and monitoring since September 11th are neither surprising nor disturbing.

So these Millennials who are used to adding yet another book to their already bulging backpacks and acquiescing to the structure and supervision placed on them appear to be receptive to advice, willing to work hard, and extremely focused on accomplishment. These are the future leaders of the Objective Force. Unfortunately, this new cohort of officers is being welcomed into an Army that is extremely supervised, highly structured, very centralized, and exceptionally busy—the wrong environment needed to transform fledgling leaders accustomed to structure into innovative, creative, out-of-the-box thinkers.

Why Johnny Can't Innovate.

The centralization of decision-making in the Army traditionally has been in the bureaucratic areas of administration, but over the years there has been a shift to increased control in the planning, execution, and assessment of small unit training. The ability to plan and conduct training at the company level has been taken away from junior officers by a system that increasingly directs the tasks to be trained, dictates the way training will be conducted, and then disrupts the training being executed. The result is an unpredictable and stifling environment of requirements, structure, and supervision that hampers most efforts toward innovation. Several factors contribute to the current leader development environment.

Too Many Good Ideas. Giving junior officers, and specifically company commanders, time to plan, schedule, execute, and assess their own company-level training should lead to increased opportunities to develop creativity and resourcefulness—necessary attributes of the Army's future leaders.[8] Company commanders can only plan training if they have discretionary time—if there are free blocks of "white space" on their calendars where they can insert their own training. A major factor in the availability of white space on the training calendar is the amount of requirements placed on company commanders from above.

For example, conventional wisdom has it that time spent on "nonmission" training has skyrocketed in the recent past. Training not directly related to the unit mission can range from Prevention of Sexual Harassment (POSH) classes to SAEDA briefings. Although the stereotypical complaint usually questions the utility of "politically correct" events such as Consideration of Others training, commanders have also shown disdain for bureaucratic or legalistic distractions such as Health Benefits Awareness briefings. A comprehensive analysis of all training requirements placed on company commanders can assess the actual impact of nonmission training on a commander's discretionary time.

Of the 365 days in the year, approximately 109 days are unavailable for training due to weekends, federal holidays, payday activities, and the Christmas half-day schedule. This results in a total of about 256 available days for company commanders to plan and execute training.[9] Requirements for mandatory training at the company level originate from Army Regulation 350-1, *Army Training*, policy letters, command training guidance, and other directives. Scrubbing all levels of command down to the Brigade level, to include Department of the Army, Major Army Command (MACOM), Corps, Division, and Installation level,[10] for anything that generates a training requirement results in the identification of over 100 distinct training requirements. Table 1 shows a partial listing of the requirements.[11]

Figure 1 depicts the distribution of training directives by the originating headquarters. Note that, as expected, most directed mission-related training requirements[12] come from Division-level or below. More importantly, most directed nonmission-related training requirements originate from DA and MACOM levels. This is critical since policy actions may be most effective in reducing the DA and MACOM requirements.

Incorporating the amount of time necessary to execute each directed training requirement (for example, training on "The Benefits of an Honorable Discharge" takes about 60 minutes a year) results in approximately 297 days of directed training.[13] Of the 297 days, about 85 percent (or 254 training days) is mission-related training and 15 percent (or 43 training days) is nonmission-related training. Two key points emerge from this analysis.

First, in the rush by higher headquarters to incorporate every good idea into training, the total number of training

20K Road March	Homosexual Conduct Policy Training
4 Mile PT Run	Individual Training Evaluation Program
Alcohol & Drug Abuse Awareness	Law of War, Geneva-Hague
Anti-Terrorism/Force Protection	Leadership Counseling Program
APFT	Live Fire Exercise
Army Continuing Education System	Military Justice
Army Core Values	Moral & Ethics Development
Army Retention Program	NBC Equipment Training
Army Safety Program	NBC, 6 Hours in MOPP
ARTEP (EXEVAL)	OCIE Inventory
Benefits of an Honorable Discharge	OMRA
Battalion Collective NBC Training	OMRP
Battalion Tactical Assault Operations	OPD/NCODP
Censorship	Operational Readiness Exercise
Change of Command	Operations Security
Child Abuse Prevention Training	Opposing Force (OPFOR)
Civil Disturbance	Organizational Effectiveness
Code of Conduct/SERE	Payday Activities
Combined Arms Live Fire Exercises	Personal Financial Readiness
Command Climate Survey	Physical Fitness
Command Information Program	12 Mile Footmarch
Command Inspection/ Organizational Inspection	Platoon EXEVAL
Command Security Program	Platoon Live Fire
Common Task Testing (CTT)	Post Run
Consideration of Others	Preparation for Overseas Replacement
Counterterrorism	Prepare for Overseas Movement
CPX	Prevention of Motor Vehicle Accidents
Crime Prevention Training	Prevention of Sexual Harassment
CTC	Protective Mask Confidence Exercise
Depleted Uranium Awareness Training	Quarterly Safety Briefs
Deployment Exercise	Rape Prevention
Division Specialized Training	Recovery Operations
Division Special Exercise	REDTRAIN
Electronic Security Briefing	Risk Assessment and Management
Emergency Deployment Readiness Exercise	SAEDA
Equal Opportunity	School of Standards
Expert Infantryman's Badge Testing	Sergeant's Time Training
Fall/Spring Clean-up	SINCGARS Sustainment Trng
Family Advocacy/Family Abuse Prevention	Special Events
First Aid	Squad EXEVAL
Fraternization	Squad LFX
Health Benefits Awareness	Stress Management and Suicide Prevention
Heat, Cold, & Hearing Injury Prevention	Suicide Risk Awareness
HIV Testing	Support of Family Members
	Survival, Evasion, Resistance
	TEWT
	Walk And Shoot
	Water Safety
	Weapons Qualification, Individual

Table 1. Partial Listing of Directed Training Requirements Placed on Company Commanders.

☒ Mission ■ Nonmission

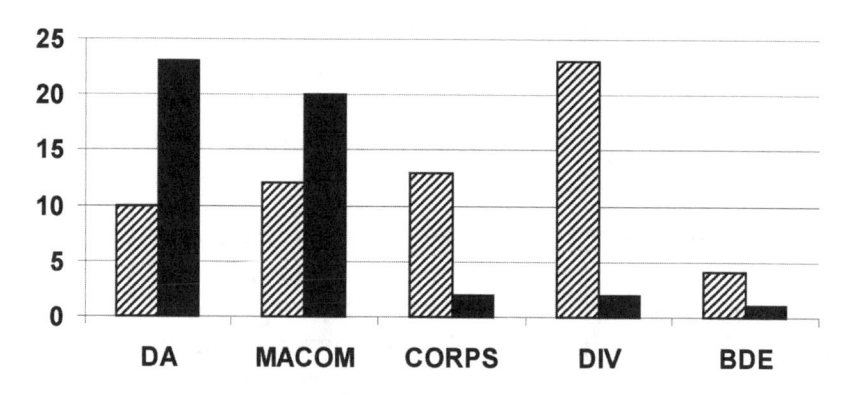

Figure 1. Number of Directed Training Requirements by Originating Headquarters.

days required by all mandatory training directives literally exceeds the number of training days available to company commanders. Company commanders somehow have to fit 297 days of mandatory requirements into 256 available training days. Second, eliminating even up to 50 percent of the nonmission-related tasks would only return at most 21 days to the company commander—still not enough time to create any white space on the training calendar. Thus, although cutting back on nonmission-related mandatory training is very appealing, the actual impact may be minimal given the relatively small quantity of mandatory training that nonmission requirements generate.

Another factor to be considered in the effort to minimize nonmission training requirements are the perceptions of nonmission training by company commanders.[14] When asked which of the nonmission requirements they would do away with, many responded to this study with observations such as, "They are all valuable. I mean, they're all important and they all need to be covered. Just not enough time on the calendar to do it all." Another commander noted, "I think the stuff that comes to mind initially [to delete] is maybe some of the required training like Consideration of Others and that kind of thing. But I'm hesitant to say that should be cut back because I think that's valuable. I think those things are valuable." Although nonmission-related training tends to be an easy target when analyzing overcrowded training calendars, most company commanders were not so willing to advocate the wholesale elimination of directed nonmission training.

An interesting question then arises when examining the directed training faced by company commanders: how are they coping with nearly 300 days of directed training in only 256 available training days? The answer is simple. Because decision makers at higher levels are either reluctant or unable to prioritize the plethora of training requirements, company commanders are forced to choose which mandatory training is executed and which is not. As one company commander noted, "You as a commander have to make the decisions on what you can and can't do and you know your unit better than anyone else, right?"

When faced by a surplus of mission and nonmission-related training, company commanders usually opt to modify the latter since the impact is less noticeable. One company commander reported, "Some units will selectively put aside that Consideration of Others training because I really need to do this training over here because I'm going to deploy or I'm going to an exercise in six months." Another commander commented, "Something has to go, so [you] pull out your values tag and say, 'Here, live by this and you will be okay,' and then cancel the training. Generally speaking, that stuff goes or gets cut in half."

If cutting back on nonmission-related training is not the solution to allowing company commanders to train and grow as leaders, then other factors that are influencing the commanders' ability to innovate must be considered. Before examining those factors, however, a key assumption must be addressed. This study assumes that giving developing officers opportunities to plan, schedule, execute, and assess their own training encourages innovation. Innovation and creativity imply the introduction of new methods, ideas, or techniques. Innovation cannot be taught in an 8-hour block of instruction. It cannot be learned over the Internet. Innovation develops when an officer is given a minimal number of parameters (e.g., task, condition, and standards) and the requisite time to plan and execute the training. Giving commanders time to create their own training develops confidence in operating within the boundaries of a higher commander's intent without constant supervision. Executing training that is planned, scheduled, and assessed by someone else develops competence, but it does not develop innovation. Reacting to the rigors and stresses of a company lane or a frenetic garrison environment encourages reflexes and off-the-cuff reactions; it does not, however, develop innovation.

Cascading Requirements. The biggest influence on the company commander's ability to plan training is the increase in directed *mission-related* training planned, scheduled, and assessed by a higher echelon. One could argue, however, that junior officers have *never* had much discretionary time. The supposed lack of junior officer freedom could be the result of over-romanticizing the past. One way to gauge the extent to which discretionary time for junior officers has been taken away is to compare a training schedule today with a training schedule from 2 decades ago. A comparison of Figure 2 and Figure 3 illustrates the increase in the quantity of directed training received from higher headquarters (and the corresponding decrease in company commander discretionary time). Figure 2 is a 12-month training schedule executed by an infantry company commander in fiscal year 1978. Black highlighting represents training planned by the company commander. Figure 3 shows another 12-month training schedule executed by an infantry company commander, except this one is for fiscal year 2001. Again, black highlighting depicts training planned by the company commander. Besides the obvious changes accompanying the introduction of PowerPoint technology, the incredible reduction in

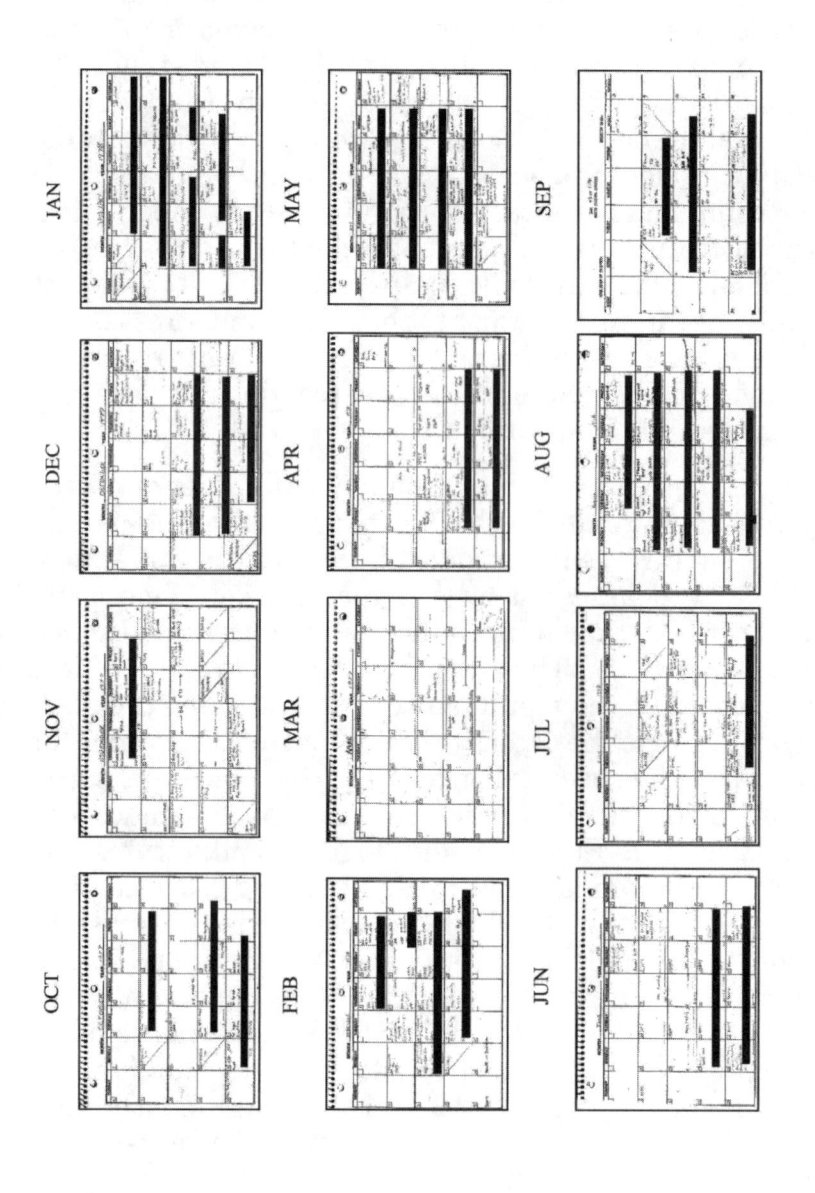

Figure 2. FY78 Company Commander 12-Month Training Calendar.

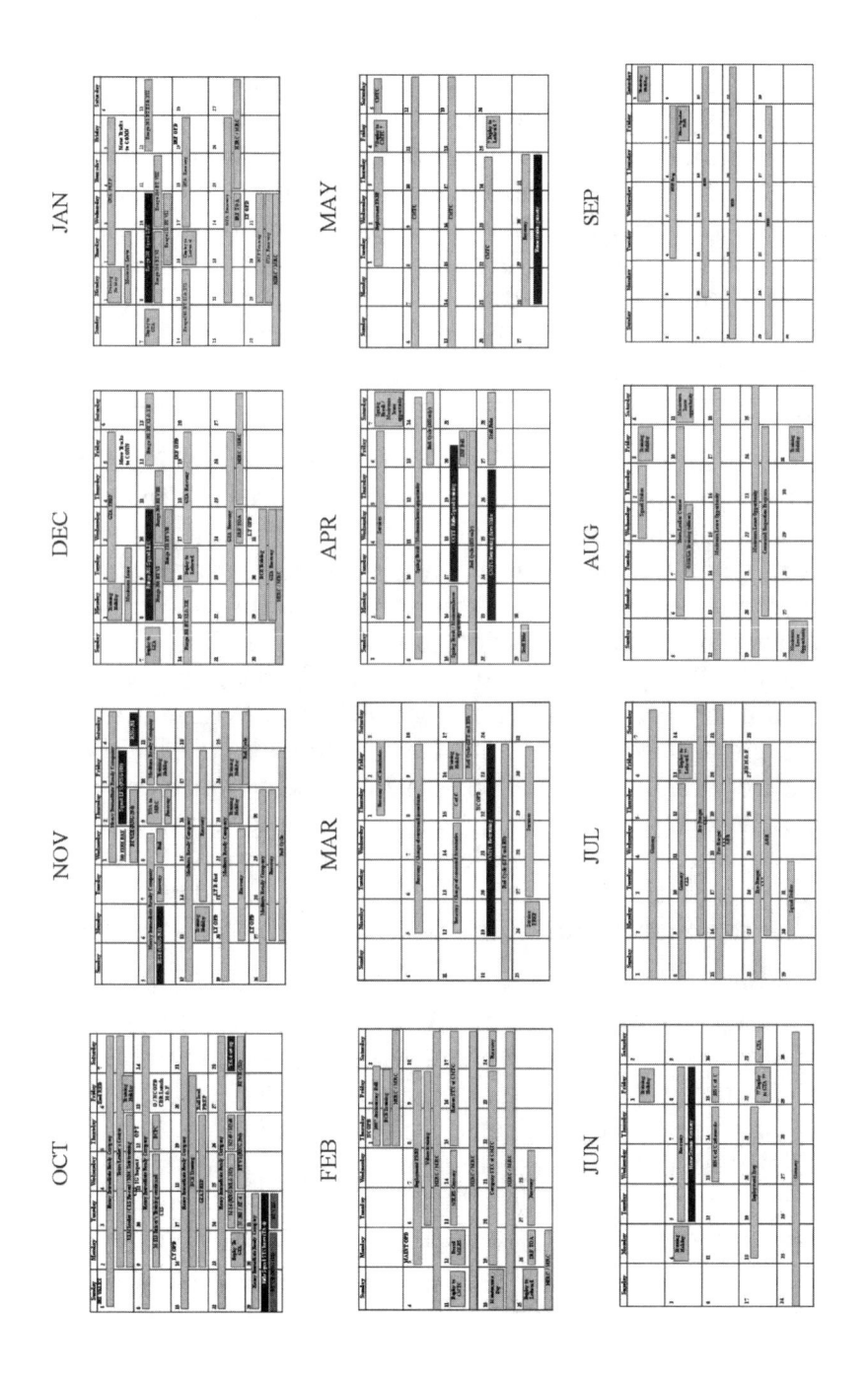

Figure 3. FY01 Company Commander 12-Month Training Calendar.

discretionary time afforded to company commanders to plan training is startling. It is important to note that the 2001 training is certainly more rigorous and complex. Nevertheless, the commander two decades ago had more time to make decisions, operate within his commander's intent, and develop confidence. While these two training calendars are only two data points, they are representative of training management in two points of time. Although the Army has made huge advances in training, leaders have inadvertently neglected to protect the company command experience as a key opportunity to develop innovation and creativity.

The influence of higher headquarters on company training increased dramatically with the start of the Army's training revolution in the mid-1970s. Along with systemizing tasks, conditions, and standards, and codifying the systems approach to training in doctrine, the training revolution included the creation of the Combat Training Centers (CTC).[15] Prior to the CTCs' existence, involvement of a company with an echelon higher than battalion during a training event was rare. Today, rotations at the CTCs with brigade operations are routine. Additionally, joint exercises involving companies have grown in frequency and scope. As a result, company commanders are much more prone to find themselves either participating in or supporting higher echelon exercises than in the past.

Participation in a higher echelon training event usually means that companies will receive the benefit of coordinating and maneuvering with higher headquarters. Unfortunately, it also means that large blocks of calendar time are taken up by higher echelon events. For example, a CTC rotation provides excellent training opportunities for a company commander. Yet the build-up to a CTC rotation will also remove about 4 months of the discretionary time from the 18 months a company commander may have in command.

The increase in higher echelon training has been accompanied by an increased demand for support to execute each training event. For example, CTC rotations typically pull company commanders and other personnel from neighboring units as augmentation or Observer/Controller (O/C) support. Another example is the Battle Command Training Program (BCTP), usually referred to as a "Warfighter," that trains division and corps staffs using computer-generated wargames. While the division or corps staffs benefit greatly from BCTPs, many subordinate units are required to provide "Pucksters" to maneuver icons representing cyber units, to provide support with admin/log functions such as tents and transportation, or to escort VIPs and retired senior officers. A typical Warfighter exercise requires 108 captains, including 80 company commanders, to serve as Pucksters or escorts for a 3-week period. One company commander, commenting on the impact of Warfighters, noted that, "Three weeks is a long time away from my company," while another said, "I'm learning nothing fighting a hex map on a computer screen."

Telling Them What and How to Do It. Not only has the discretionary time on a company commander's training calendar been displaced by higher echelon directives, but *what* will be trained and *how* it will be trained has also been increasingly dictated by higher headquarters. Dictated training serves to take away a company commander's initiative by requiring commanders essentially to follow a script during training. Instead of planning and developing training from the bottom up, as originally envisioned by FM 25-100, *Training the Force,* and FM 25-101, *Battle Focused Training,* higher echelons are determining what will be trained and how it will be executed. Consequently, company commanders have little ability in targeting training to their analysis of their units' weaknesses.

One reason for the shift from a bottom-up to a top-down approach to training was the emergence of real world deployments that went beyond existing Army doctrine. With the absence of established doctrine to prepare units for operations other than war, lessons learned from the first units deployed were codified and passed to subsequent units. Mission Rehearsal Exercises (MREs) were created to merge the lessons learned from previous deployments with the idiosyncrasies of a unit's pending deployment. Detailed training checklists are now used by higher echelons to conduct company and platoon training lanes that provide realistic, practical training. Table 2 illustrates the rigorous training regimen presented to company commanders prior to a MRE. The training is comprehensive and thorough, yet it removes all discretion from the company commander.

Individual Training
- Country Orientation
- Media Awareness
- Rules of Engagement
- Environmental Threats
- Convoy/Driving Hazards
- Mine Awareness/Countermine
- Force Protection
- JMC Handbook Training

Platoon Lanes
- Apply the rules of engagement
- Conduct a weapons/equip seizure
- Conduct convoy escort
- Conduct dismounted/mounted patrols
- Conduct link-up with escorted convoy
- Conduct patrol
- Conduct WSS inspection
- Develop and communicate a plan
- Establish a hasty CP
- Liaison with civilians
- Liaison with EAF
- Liaison with local police
- Liaison with local police and IPTF
- Prepare for patrolling
- Protect the force
- React to a civilian with a weapon
- React to civilian vehicle accident
- React to civilians
- React to hostile civilians
- React to human remains scene
- React to indiscriminate fire
- React to Media
- React to PIFWC
- React to police CP
- React to UXO

Company Lanes
- Block Hostile Civilians
- Conduct a Vehicle Escort
- Conduct BILAT w/ Mayor & Police
- Conduct Camp Security Ops
  - Personnel Access Point
  - Vehicle Access Point
  - Guard Towers
  - QRF
  - Change of Guard
- Conduct a WSS Inspection
- Conduct Patrols
- Enforce FOM/Security for Local Elections
- Establish an OP
- Establish Checkpoints
- Establish Retrans w/ Security
- Extract PDSS
- Gather Intelligence
- LNO w/ police, IPTF, & Mayor
- LNO with EAF
- React to UXO
- Respond to Hostile Civilians
- Respond to Police Check Points (Legal & Illegal)
- Secure/Guard Remote Site

Table 2. Partial Listing of SFPR/KFOR Deployment Training Planned by Higher Echelons.

In addition to the planning responsibility being taken over by higher echelons, the assessment function in the training management cycle has also shifted upwards. In an effort to minimize risk, higher headquarters are not only responsible for observing and assessing training, but they must also "certify" units as ready. Certification evaluates the training against established standards—a necessary step in the training management cycle per FM 25-100. What has changed over the years, however, is that this role has migrated up the chain of command. Instead of company or battalion commanders evaluating small unit level training, deployment plans direct things like, "All training conducted ... will be certified by the first O-6 in the chain of command. He/She will certify that the training was conducted to standard prior to deployment."[16]

An unintended by-product of the centralization of the planning, scheduling, and assessment functions is that commanders are removed from the training management process and instead are responsible only for moving their units through the training lanes. While the training process works toward attempting to guarantee readiness, leader development is sacrificed. A battalion commander on a Kosovo rotation astutely noted this tradeoff during a four-month Kosovo train-up and stated,

> The only training I planned and conducted [was] innovative tank gunnery, some OPDs, and NCODPs. Other than that, all the training was planned, resourced, and analyzed by my higher HQs ... Regardless, the training met the standards necessary to produce soldiers that were well-trained and prepared to accept the hazards of duty in Kosovo.

It is understandable that a centralized, risk-minimizing training approach is used to prepare for a real world contingency, but similar templates and gate strategies are used for CTC train-up periods. For example, FORSCOM Regulation 350-50-1 details what training will be executed prior to a rotation at the National Training Center. Higher headquarters plan what, how, when, and where the training will be conducted. Battalions prepare squad and platoon lanes, while brigades plan and resource company lanes. In many cases, the first time a company commander has a company on line is during an evaluation by an external O/C team.

The upward shift of responsibility for the planning of training is ironic given the original intent of the current training doctrine. In the foreword to FM 25-101, General Carl Vuono wrote,

> While senior leaders determine the direction and goals of training, it is the officers and noncommissioned officers at battalion, company, and platoon level who ensure that every training activity is well planned and rigorously executed. This manual is for them—the leaders at battalion level and below.[17]

Realistically speaking, leaders at battalion level and below do not need FM 25-101 simply because nearly all the training they encounter is directed and planned by higher echelons.

It is paradoxical that the Army now conducts the most effective training in the world—training that produces confident, competent soldiers—yet the system to create that training has become incredibly centralized. It is tempting to rely on field training to produce innovation and creativity in spite of the over-structured garrison experience, but as Major General (then Lieutenant Colonel) James Dubik pointed out nearly a decade ago,

> A unit cannot operate centralized in garrison and decentralized in the field. A commander is mistaken if he believes that such a conceptual shift is possible. Subordinates who, in garrison, are used to deferring decisions until consulting with, and receiving approval from the battalion commander will not suddenly be able or willing to make the judgments required of them in training or in combat.[18]

One Chance to Train. Two additional factors encourage higher echelon trainers to pull the planning, scheduling, and assessing training functions upward—limited resources and officer competition. Limited resources such as land, ammunition, spare parts, time, and fuel all compel a higher headquarters to allocate the scarce resources and then ensure efficient use of resources by taking over the planning function. In an austere training resource environment, higher-level control of planning and scheduling seeks to minimize the chances that resources will be less than optimally expended. Without the luxury of enough resources to allow re-training or the possibility of "wasting" a training resource due to lower-level mistakes, higher-level commanders step in to maximize efficiency and short-term effectiveness.

For example, local training areas are often a limited resource, especially in Europe. Because of limited range time, any time to train is especially valuable. As one company commander noted, "There was extreme pressure for us to hurry up and get off the range . . . because you don't want to *range sponge*. You don't want to use up range time." To minimize the possibility of ineffective training consuming the limited resources such as range time, higher-level commanders are apt to plan and schedule the training for the company and also require certification of all tasks leading up to the training event.

Officer competition also increases the tendency for higher-level commanders to subsume the company commander's training management cycle. Despite the assurances that training is not *testing*, there will always be a degree of leader evaluation involved with every training event. Impressions of leaders are sure to be influenced by the relative "success" of their subordinate commanders' training. Beating the OPFOR becomes more important than identifying training shortcomings; winning becomes more important than learning. Higher echelon commanders can help increase the chances for success by leveraging the experience and breadth of knowledge found in higher echelon trainers. While the probability of success increases and the probability of wasted resources decreases with training planned at levels higher than the company, it directly diminishes the ability of the company commander to innovate.

The Futility of Planning Training.

In addition to directing what will be trained, when it will be done, and how it will be conducted, higher echelons are also disrupting actual training being executed through the damaging effects of late-notice taskings. Late-notice taskings rob a company commander of predictability as planned training is either canceled or degraded. One commander noted, "I can tell you that almost every week there is a change on the training schedule due to unpredictable missions or taskings."

Taskings may include supporting other units during training by providing O/C support; providing soldiers to fill in for infrastructure shortfalls on post by serving as school bus drivers or sorting metal at the recycling center; or supporting the community by moving desks in off-post elementary schools partnered with the unit. It is not so much the taskings in themselves that disrupt training; instead it is the late receipt of the taskings that puts commanders in a reactive mode.

To sample the effects of taskings on company commanders, two CONUS posts were analyzed for this study. All taskings arriving at the Post level were examined during a 90-day period. Taskings included training support, nonmission-related support, and community support requirements. At the Post level, it was found that over 80 percent of the requirements reviewed were submitted for tasking support more than 2 months prior to execution. In other words, the vast majority of taskings arriving at Post level gave the Post at least 8 weeks to respond—a reasonable timeframe. Of the 80 percent of the taskings arriving on time at Post level, however, 76 percent arrived at the company level inside the FM 25-100 recommended 6-week training calendar lock-in. As a result, training planned and reflected on the training schedule was disrupted in some manner by a large majority of the taskings.

Red, Amber, and Green cycles often serve to soften the impact of short-notice taskings since units in the Red cycle are prepared for short-fuse requirements. Unfortunately, large parts of the Army, especially those OCONUS, do not follow the Red, Amber, Green cycle rotation. Additionally, many commanders noted that short-notice taskings, specifically those tasking training or deployment support to another unit, still intrude upon Green cycle training time.

The Administrative Burden. Administrative tasks have always been and probably always will be an undesirable, yet necessary part of command. Nevertheless, today's company commanders are under a heavier administrative burden than commanders in the past. Advances in technology combined with a growing desire to avoid uncertainty have created a culture where senior leaders constantly pump company commanders for information and get involved in company level leadership. Today's company commanders are busy, just like commanders in the past, but much more of their energy is spent reacting to administrative requirements.

For example, tracking data has become a large part of a company commander's daily routine. Commanders are tracking items ranging from personnel awards to unit level maintenance training using a variety of tracking matrices. Many company commanders interviewed for this study arrived with thick "Smart Books" crammed full of statistics and information on their unit and soldiers. How much data are commanders collecting? Table 3 helps to illustrate the current situation by listing over 125 examples of data that company commanders report they are recording.

Data tracking requirements originate from three sources. The first and most obvious source is regulations. Examples include completion of monthly sensitive item inventories, sexual responsibility training, and family care packets. The second source of requirements is command

Sensitive Item Inventory	Arms Room Access	EFMP
10 percent Inventory	Key Control Roster	Hearing Conservation
Overweight Program	Drown Proofing	Leaves & Passes
Pregnant Soldier	Anti-Terrorism	Personnel Actions
APFT Failure	Law of War	UMR
PLL Review	AWOL Baggage Control	USR Input
ULLS-G	Alcohol & Drug Prevention	Congressional Inquiries
ULLS-A	Environment Awareness	Unit Funds
Automation Certification	NEO	NBC Calibration
Range Officer	Extremism	Hand Tool Calibration
Drivers Licensing	Sexual Harassment	AUSA Membership
Master Drivers Training	Depleted Uranium	Division Association
Unit Level Maintenance	Military Justice Brief	Urinalysis Program
Night Vision Device	TCACCIS	Force Protection
Homosexual Policy	AER Brief	College Attendance
Sexual Awareness	CFC Drive	PMCS DA 5988
Fraternization Policy	Blood Drive	Parts Not Installed
EO Council	Jr Officer Development	Hometown News
Demographics	Jr Officer Certification	Funeral Detail
CO2 Training	Distinguished Leader	Small Arms Repair Parts
CO2 Trainer Cert	Jr EM Counseling	Newsletter Input
Re-Enlistment	NCO Counseling	Lautenberg Amendment
Family Support Group	PERSTAT	Volunteerism
AFTB	PERSTEMPO	Pre-Chg of Cmd Inv
Family Care Plans	Unit Climate Survey	Reports of Survey
Financial Management	Gains	IARs PAI CSDP
Green-to-Gold College	Sponsorship Letters	CBI Inspections
CLS Certification	Awards	Ethics Brief
CLS Bag Certification	Schools Requirements	Army Values
HAZMAT Qualification	NCOERS	Long Weekend Safety
MILES Certification	OERS	POV Inspections
Cdrs Finance Report	Rating Schemes	Facility Work Orders
Promotion Worksheets	Non-Deployables	Income Tax Brief
Safety Briefs	Adverse Action	NCO Tattoo Inspections
BOSS	Flags	Unit Load Plans
Alert Roster Posting	Bars	Family Counseling
SAEDA & OPSEC	Medical Records Check	Physical Eval Board
PDP	MOS Shortages	Savings Bond Drives
SRP Packets	Physical Profile Review	Voting Assistance
Crime Prevention	30 Day Arrivals	Tax Assistance
	90 Day Loss Separations	Payday Activities
	BMM & TD	IMPAC Card
		Suicide Prevention
		Fire Marshalls
		Risk Assessment

Table 3. Statistics Collected by Company Commanders.

directed. Examples range from community volunteer hours to soldiers who have vehicles with Firestone tires. The third source of data collection stems from the need to prove certification. Consideration of others facilitator training, unit demographics for EO council selection, and range officer certification are examples of data that need to be recorded and monitored.

Several factors exacerbate the data tracking role of a company commander. Company clerks and training NCOs largely have been eliminated, yet automated databases still have to be updated and queries answered. As a result, company commanders are often personally involved in storing and retrieving data. Additionally, many CSS units do not have company executive officers, further pushing the administrative burden onto the commander. Company commanders today are extremely busy, but they are often busy with administrative tasks instead of training their unit. One company commander noted, "My biggest problem is that I fell more like an administrator than a commander. ...All I can say is that I honestly feel that the Army is not about leading troops anymore. It is about numbers. It is about compiling information about those troops." Another commander added, "We spend a lot of time gathering information. That is one of my complaints . . . Succeeding in the Army does not seem so much about leading soldiers, it is about requiring data on soldiers."

A major influence in the increased administrative burden on company commanders has been the impact of e-mail. E-mail allows staffs and higher-level commanders to bypass lower-level staffs and directly query commanders for information. The diminished ability of undermanned and inexperienced staffs in the force combined with the speed of automation has greatly increased the reliance on e-mail. As one company commander put it, "Thanks to e-mail, the most trivial question that any senior leader can come up with can be asked in a matter of seconds. Staff work and staff analysis is a lost art."

It is important to note that there is nothing inherently harmful about e-mail—most commanders like e-mail since it keeps them connected with their peers. It only becomes detrimental when it is used to circumvent staffs and add to the administrative burden of company commanders. Many of the company commanders interviewed resembled action officers in the Pentagon, arriving early in the morning to see what taskings were in their in-box, logging on again at lunch to answer any queries, and then checking again at night before heading home.

The Garrison Template. Occasionally, company commanders may receive a week of discretionary time to plan their own training. Unfortunately, most of the Army now places an administrative template over training time while in garrison. Figure 4 shows an actual template similar to those used throughout the Army. Mondays are reserved for command maintenance in the mornings and the afternoons are filled with meetings at battalion or brigade level. Fridays are usually spent preparing for the next week or for special events such as compensatory time or payday activities. Thursday mornings are occupied by sergeant's time, followed by family time in the afternoon. The result is that Tuesdays and Wednesdays are the only days available for any company commander-generated training. As one commander pointed out,

> It seems that all of the higher driven events, training meetings, command and staff, maintenance meetings, sergeant's time, motor stables, and events like that are blotched throughout the week in such a way that you couldn't actually sit down and conduct other training in between.

The administrative template is a good example of the overall issue of over-control. Every directed activity, from motor stables to family time, is a good idea when analyzed in isolation. Put all the directed requirements together, however, and the life of a company commander is spent executing somebody else's good ideas. Individually, directed training requirements make sense and benefit soldiers and

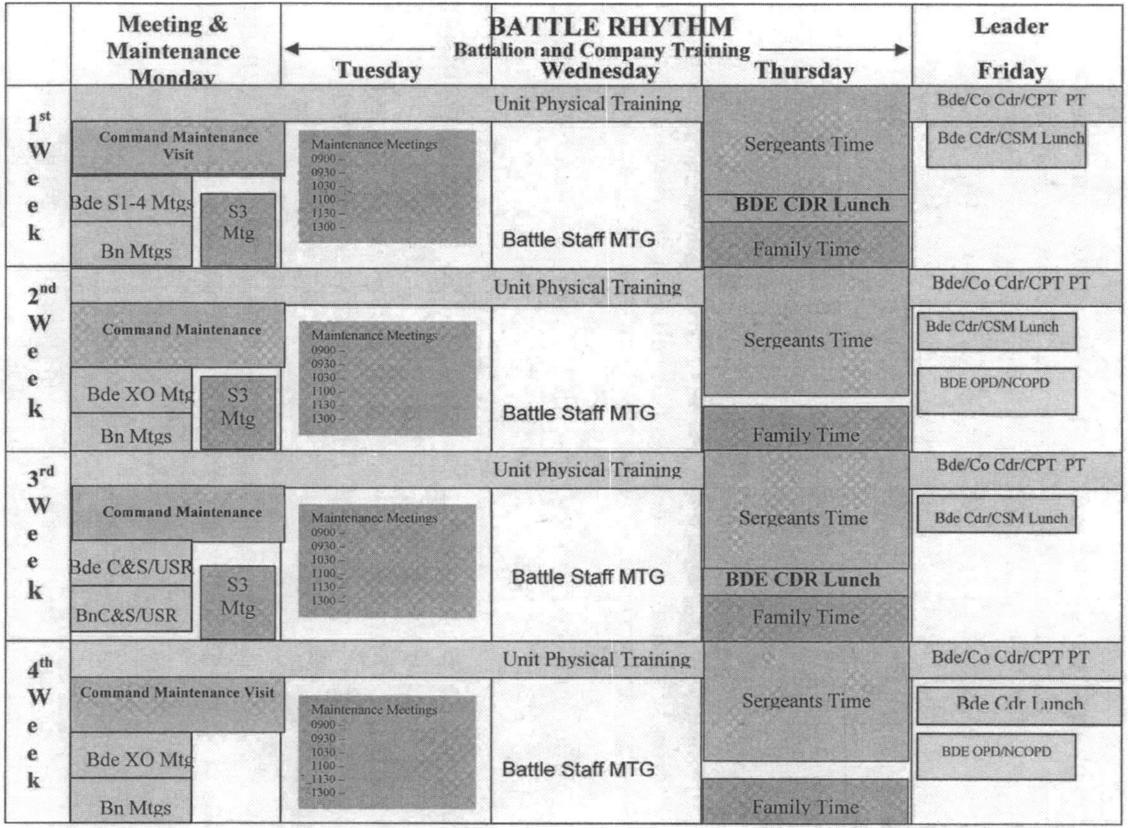

	Meeting & Maintenance Monday	Tuesday	BATTLE RHYTHM Battalion and Company Training Wednesday	Thursday	Leader Friday
1st Week	Command Maintenance Visit / Bde S1-4 Mtgs / S3 Mtg / Bn Mtgs	Maintenance Meetings 0900 – 0930 – 1030 – 1100 – 1130 – 1300	Unit Physical Training / Battle Staff MTG	Sergeants Time / BDE CDR Lunch / Family Time	Bde/Co Cdr/CPT PT / Bde Cdr/CSM Lunch
2nd Week	Command Maintenance / Bde XO Mtg / S3 Mtg / Bn Mtgs	Maintenance Meetings 0900 – 0930 – 1030 – 1100 – 1130 – 1300	Unit Physical Training / Battle Staff MTG	Sergeants Time / Family Time	Bde/Co Cdr/CPT PT / Bde Cdr/CSM Lunch / BDE OPD/NCOPD
3rd Week	Command Maintenance / Bde C&S/USR / S3 Mtg / BnC&S/USR	Maintenance Meetings 0900 – 0930 – 1030 – 1100 – 1130 – 1300	Unit Physical Training / Battle Staff MTG	Sergeants Time / BDE CDR Lunch / Family Time	Bde/Co Cdr/CPT PT / Bde Cdr/CSM Lunch
4th Week	Command Maintenance Visit / Bde XO Mtg / Bn Mtgs	Maintenance Meetings 0900 – 0930 – 1030 – 1100 – 1130 – 1300	Unit Physical Training / Battle Staff MTG	Sergeants Time / Family Time	Bde/Co Cdr/CPT PT / Bde Cdr Lunch / BDE OPD/NCOPD

Figure 4. Administrative Template for Garrison.

leaders. In the aggregate, they produce an environment that stifles initiative and develops leaders reliant on structure, not their own judgment.

The Training Façade. Although all the components of the model discussed so far have impacted on innovation, there is also another key by-product of the factors analyzed thus far—the training management system has largely become moot. Company training meetings are still conducted. Company training schedules are still posted. Quarterly training briefs are still briefed. Yet, when company commanders were asked how valid their training schedules were, specifically the 6-week lock-in, they responded with comments such as,

> My battalion commander fought [for the 6-week lock-in] when he came in. He was unsuccessful. He has now said, 'We will forecast training to 6 weeks, and we will attempt to lock-in at 4.' And, like I said earlier, very rarely do we get a full week lock-in.

Another commander stated, "There's a 6-week lock-in. But it almost never works. In fact I typically don't know what's going to happen the next day." Finally, one commander's comments pointed to the growing cynicism about training management and stated, "The 6-week training calendar is a joke. It is not even required to be signed until 3 weeks out. That is a division standard here and they know that."

Many senior leaders understand that except for training directed by higher echelons, training schedules are largely nonbinding. One company commander stated,

> We had this discussion in our training meetings with commanders and the colonel. He understands our frustration, but we are getting beat up to turn in training schedules 6 weeks out, and we are just like—Why? It is going to change, it always changes.

A training façade emerges when captains at the career courses are taught how to plan company training per FM

25-100 and -101, yet discover when they take command that there are few or no opportunities to plan training. The façade grows when commanders cancel or downsize directed nonmission-related training to accommodate overloaded training calendars, but such actions receive only a wink by higher levels. The façade is reinforced when commanders are asked to plan and brief training schedules, fully knowing that late notice taskings will probably invalidate much of what they are briefing. Finally, the façade spreads to the entire unit when troops have to wait until morning formation to learn what is really going to happen that day.

Conclusions.

The situation in which the Army finds itself is oddly paradoxical. Future leaders should be adept at operating in unstructured, ambiguous environments, yet the Army is relying on a centralized, over-structured system to provide that capability. As a result, an entire cohort of junior officers is inadvertently being produced whose company command experience consists mainly of responding to directions and disruptions from higher headquarters. Discretionary time has been replaced by sergeant's time, innovating has been replaced by reacting, and creativity has been replaced by certification.

The Army's changing approach to training parallels a similar shift in how the Army views leader development. On one hand, training management can be a continuous cycle of Mission Essential Task List development, identification of strengths and weaknesses, planning, execution, and then assessment. This approach implies development, mistakes, corrections, and growth. The other perspective—and this is the direction the Army has moved toward—trains for a specific situation, mission, or predetermined objective. The focus is on accomplishment, not learning. Likewise, the company command experience can be viewed as an operational assignment serving the purpose of developing leaders. This approach correlates a successful command with the degree of development and growth in the leader. The other perspective views company command as an opportunity to execute a mission. Success is no longer gauged by how the commander has progressed in development as a leader, but instead focuses on what the commander managed to accomplish while in command. If the company command experience is to remain a key part of the leader development system, then there must be a corresponding shift within the training management system back to a developmental approach.

Two important facets need to be addressed before going further. First, while *micromanagement* is certainly associated with the issues discussed thus far, it is not the main focus of this monograph. Micromanagement implies that leaders are getting too involved in the business of their subordinates. To a large extent, the move toward over-centralization reflects that. But putting the blame on micromanagement views the situation too narrowly. The current situation of over-control reflects the *culture*—not just the leaders of the Army. Most company commanders do not believe their battalion commanders are the reason for so much over-control. They view them as merely passing on the requirements and directives from higher headquarters. The Army now has a culture where the obsession with minimizing risk and uncertainty has pervaded not just the leadership, but also the way the entire institution thinks and works. The Army has moved to the point where the current degree of structure, control, and centralization is accepted as proper and necessary—which leads to the second point.

Because company commanders have not experienced anything other than the current environment in the Army, *they are largely unaware of any changes in the culture.* There has not been a huge outpouring of discontent about the paucity of discretion afforded to company commanders. As far as most company commanders can tell, they are accomplishing the mission and taking care of soldiers just as company commanders have been doing for decades. Only now, accomplishing the mission consists of executing whatever is directed from higher headquarters, not creating a workable plan from the commander's intent. Taking care of soldiers still consists of ensuring soldier well-being, but now also includes collecting a myriad of statistics on them to report to higher headquarters.

One unique group that shed some interesting highlights on the company command experience were the commanders interviewed while deployed in Kosovo. These company commanders were in an environment where the only guidance from above was broad and vague. Meetings with higher headquarters were restricted to a bi-weekly frequency and higher commanders spent most of their time giving their intent and supporting, not directing, company activities. These company commanders enjoyed their autonomy, responsibility, and ability to innovate within the commander's intent. Interestingly, they looked forward to redeploying to see their families, but dreaded falling under the centralized, controlling environment upon return to home station. These commanders, however, are in the minority. Most company commanders only occasionally experience the exhilarating rush of responsibility that comes with commanding in an ambiguous, unsupervised environment.

As discussed earlier, the next generation of officers entering the Army is accustomed to being supervised and structured—that is what they grew up with. Yet during any discussion of transforming the Army, it is assumed that the Army's structured, centralized system will take these easily-structured, easily-supervised officers and develop them into out-of-the-box thinkers needing little structure or supervision. The probability of that happening is slim unless changes are made.

Changing an Army.

While it is tempting to think that policy actions can remedy the situation presented thus far, there are two reasons why policy changes alone will not be effective. First, the push for over-centralization and over-control has permeated the entire Army. Leaders at all levels unconsciously reinforce it, as do staff officers throughout the Army. It is reflected in the everyday way business is done, ranging from mandating 4-day holiday weekends (instead of leaving the decision up to local commanders) to the complex process of filing a travel voucher. A pure policy approach would inevitably be too narrow and temporary in restoring opportunities for innovation for junior officers.

Second, policy fixes alone will probably aggravate, not mitigate the situation. Directing that company commanders must have a week of training calendar white space every quarter will appear to help, but unless the overall culture changes, company commanders will still have to cram the same amount of requirements into the remaining 48 weeks of the year. Innovation will continue to suffer unless the pressure is lifted off company commanders. Likewise, mandating that no new requirements can be placed on company commanders unless approved by the first general officer in the chain of command trims back on requirements, but inadvertently contributes to the trend toward centralization that is the root cause of the problem.

Thus, revitalizing the company command experience to one where creativity and innovation can flourish will require more than just policy changes. It will require a change in the way the Army approaches problems and issues. It will require changing the Army's culture to one where subordinates are free to innovate. Change must start at the top and then eventually work its way down to the bottom. One cautionary note, however—if the Army chooses to embark upon this cultural change, it must see it through to completion. With a training façade already existing, announcing another "Power Down" initiative and then not following through with it may make the current situation even worse.

A difficult aspect of this cultural change is that, in one sense, it requires senior leaders to do *less*, not more. Senior leaders (colonels and higher) need to be convinced to give standards, some basic guidelines, and then let the commanders train.

Creating a Culture of Innovation
- Invest in leaders, not the system.
- Learn to live with uncertainty.
- Protect company commanders from external distruptions.

Leaders have to overcome the desire to tell subordinates how to do it, continually check on it, grade it by an external O/C, and then certify it by the first O-6 in the chain of command. Subordinate leaders can be given tasks, conditions, and standards, as well as any parameters, but senior leaders must refrain from detailing *how* a task is to be accomplished. Senior leaders should demand *a* solution, not *the* solution. Leaders will have to learn to live with the uncertainty of not knowing or controlling everything.

The Army has to ask itself what echelon levels rate the top priority—if it is Corps or Division, then they need to be exercised often. But if future battles will be fought by leaders at brigade, battalion and below, then these leaders need to be protected as they train as company commanders today. Similarly, if the immediacy of the real world demands extra control from above, then ensured readiness must be artfully balanced with subordinate leader development. Without the deliberate relinquishing of control, exceptions will begin to proliferate and the desired cultural change will not occur.

Although transforming the Army's culture to one where subordinates are free to innovate will require senior leaders to provide less detailed direction and structure, it will actually require more effort from the senior leaders themselves. Senior leaders will have to increase the time and energy spent on developing and improving subordinate leaders—not the system of procedures, checklists, or "The Model." Senior leaders will not only have to shift much training back down to battalion and company level, but must also respect the company commanders' time by protecting them from short-notice taskings and changes. The uncertainty that results from relinquishing control to lower levels will have to be absorbed by senior leaders themselves, not by requiring constant supervision or data collection. If reliance on statistics, metrics, and briefings to ensure personnel and unit readiness is relaxed, then leaders will have to increase personally checking and monitoring attainment of training standards.

Another difficult part of this cultural change is that the current culture appears to satisfy the needs of today's Army. Deployments to Bosnia and Kosovo have gone incredibly well. SOF forces and lightfighters are accomplishing the mission in Afghanistan and other locations. Readiness levels are high and the Army continues to execute its duties efficiently and effectively. While junior officer attrition remains a problem, the Army as a profession is still attractive to most junior officers. Yet the cultural change is not for today's situation; it is for the future Army. In order for the Army to be a relevant force in the future, transformation to the Objective Force must occur—and that includes transforming the way leaders are developed.

Changing the Army's approach to leader development will not be easy. Objections will come from three avenues of approach. First, the leader development process has served the Army well for the last two hundred years. Leaders will say, "Why should we change it now? The American soldier is historically famous for being innovative and creative; hold steady and don't panic." This approach denies the transformation vision. That vision calls for transforming not only the modernization dimension, but also the human dimension. Objective Force battles cannot be fought with Legacy leadership. To insist that the current leader development process is sufficient ignores the changed world of the future.

The second objection will agree that change is needed, but it is the junior officers who need to change, not the culture. This approach places the cause of the leader development mismatch on the immaturity, lack of experience, wrong expectations, or misguided perceptions of junior officers. If company commanders could only get more time in command; if they could only receive more guidance; if they could only see the whole picture; then this problem would be solved. This objection ignores the environment that currently exists. It ignores the stifling set of rules, requirements, and disruptions that steal any amount of initiative in the current company command experience of junior officers.

This objection actually hinges on the level of trust between senior and junior officers. The current system replaces trust in junior officers with a myriad of controls, checks, and constraints. Lifting this environment of over-control, according to this objection, will result in anarchy. Company commanders will fill their training schedules with white water rafting and meaningless training. Ironically, the 5 percent of commanders, who might display bad judgment if given that freedom, are today protected by the security of a system whose structure prevents any judgment—good or bad.

The final objection will argue that while nearly all of the training for company commanders is directed and dictated, commanders can still innovate within those boundaries. This approach points out that there is plenty of innovation and creativity being developed as company commanders escort their companies to STX lanes or follow a prescribed menu of training. According to this view, innovation can be developed in other ways besides getting the opportunity to plan training. In a sense, this objection has some merit. Innovation can be developed when company commanders attempt variations on how to execute a task after being told what and how to do it. This innovation, however, is reactive in nature, not proactive. It is innovation, but a limited type of innovation and not the same innovation needed in leaders of the Objective Force.

Changing a culture is extremely difficult. Derailing a cultural change, however, is fairly easy. When the proposed cultural change occurs, subordinate leaders will begin to exercise more judgment. In some cases, poor judgment (and learning) will occur. The reaction of senior leaders, both internal and external to the Army, will be crucial to the success of the cultural change. An unforgiving reaction could stall any attempts at changing the culture. Additionally, the Army will be able to create a culture of innovation only if all segments and levels of the Army participate. With the current personnel system moving junior leaders geographically and up the rank structure every few years, an isolated cultural change at a particular location or level will die out quickly.

This monograph is about the Army's success as a learning organization. As an organization, the Army has benefited from the lessons learned "in the box" at the Combat Training Centers, the best techniques from recent deployments, and the wisdom of past commanders who have "been there and done that." As an organization, the Army continues to learn and codify the best practices in all arenas. Yet the zeal for *organizational* learning has been at the expense of *individual* learning. If future battalion and brigade commanders are expected to be innovative, independent thinkers, then the Army must change the way it does business in order to give today's junior officers the opportunity to innovate and think independently now. With a new generation of officers entering the profession, the Army transformation is the perfect time for that change.

ENDNOTES

1. Fort Lewis Interim Brigade Combat Team website, *http://www.lewis.army.mil/transformation/faqs.htm*, May 2001.

2. Lieutenant General William M. Steele and Lieutenant Colonel Robert P. Walters, Jr., "21st Century Leadership Competencies," *Army Magazine*, August 2001, p. 29.

3. General Wesley Clark, "Implementing Needed Change—Reforms to Structure and Process," in *National Strategies and Capabilities for a Changing World*, Washington: Institute for Foreign Policy Analysis, 2000, p. 72.

4. General Frederick J. Kroesen, "Korean War Lessons," *Army Magazine*, November 2000, p. 33.

5. For a look at Generation X officers, see Leonard Wong, *Generations Apart: Xers and Boomers in the Officer Corps*, Carlisle, PA: Strategic Studies Institute, 2000.

6. This summary comes largely from Neil Howe and William Strauss, *Millennials Rising: The Next Great Generation*, Random House: New York, 2000. Also see David Brooks, "The Organization Kid," in *The Atlantic Monthly,* April 2001, pp. 40-54 for an interesting view of the Millennial Generation.

7. Dana Milbank, "Bush Tour Will Promote National Service," *Washington Post*, March 4, 2002, p. A2.

8. The genesis of this study was a Chief of Staff of the Army charter to determine all directed training requirements placed on company commanders and then to recommend the elimination or modification of 50 percent of the compliance (or nonmission) tasks faced by commanders. The rationale for eliminating 50 percent of nonmission-related training requirements was the belief that company commanders had too little discretionary time to plan their own training—no "white space" on their training calendars—because of all the mandatory training requirements directed at company commanders.

9. This estimate is extremely conservative since it does not include 4-day weekends; assumes that training can be conducted on 8 out of 52 weekends and half the paydays; and only assumes 5 days for Christmas half-day schedule. Training Circular 25-8 gives 242 vice 256 days as an Army baseline goal for annual training days.

10. To gather training requirements, four MACOMs, three Corps, and eight Divisions were visited and analyzed.

11. Table 1 is based on a combat arms company in either CONUS or OCONUS.

12. Mission-related requirements are tasks that directly relate to the unit's combat mission, e.g., maintenance or marksmanship, while

nonmission-related requirements contribute, but do not directly relate to the unit's combat mission, e.g., Equal Opportunity training.

13. A duty day is assumed to be 10 hours. Estimates for the duration of each training requirement were determined using information from the Center for Army Lessons Learned and a panel of former battalion commanders.

14. Over 180 company commanders were interviewed individually using a structured interview procedure. Interviews were audio taped and transcribed for analysis. The interviews concentrated on company commander perceptions of the training and leader development environment. Interviews were conducted throughout the active Army to include deployed commanders. The company commander sample was 44 percent combat arms, 28 percent combat support, and 28 percent combat service support. Average time in command for interviewed company commanders was 11.4 months.

15. Lieutenant General Thomas N. Burnette, Jr., "The Army's Second Training Revolution," *Army Magazine*, October 1997, p. 114.

16. SFOR Deployment Directive, Annex T to SUPPLAN 3.

17. Headquarters, Department of the Army, Field Manual 25-101, *Battle Focused Training*, Washington, DC: U.S. Government Printing Office, 1993.

18. Lieutenant Colonel James M. Dubik, "Decentralized Command: Translating Theory into Practice," *Military Review*, June 1992, p. 37.